Jadon, Way Up High in the Apple Tree

By Carol Van Zanden

Way up high in the apple tree.
4 little apples smiled down at me.
I climbed that tree as high as I could.
I picked apples for my family.
They were very good!

Grampa Van, Larkin does not like apples.

I will try an apple.

This apple is GOOD!!

Would you like to pick some apples to share with your family?

I would like to pick apples "way up high in the apple tree" to share with my family.

I could pick some apples if I climb on the fence.

Hang on to the branches, Jadon.

See how high I am picking apples

from "way up high in the apple tree."

I have 4 apples to put in a box.

The apples look GOOD, Grampa Van.

These are BIG apples to share with my family.

Thank you, Grampa Van for helping me pick apples from "way up high in the apple tree."

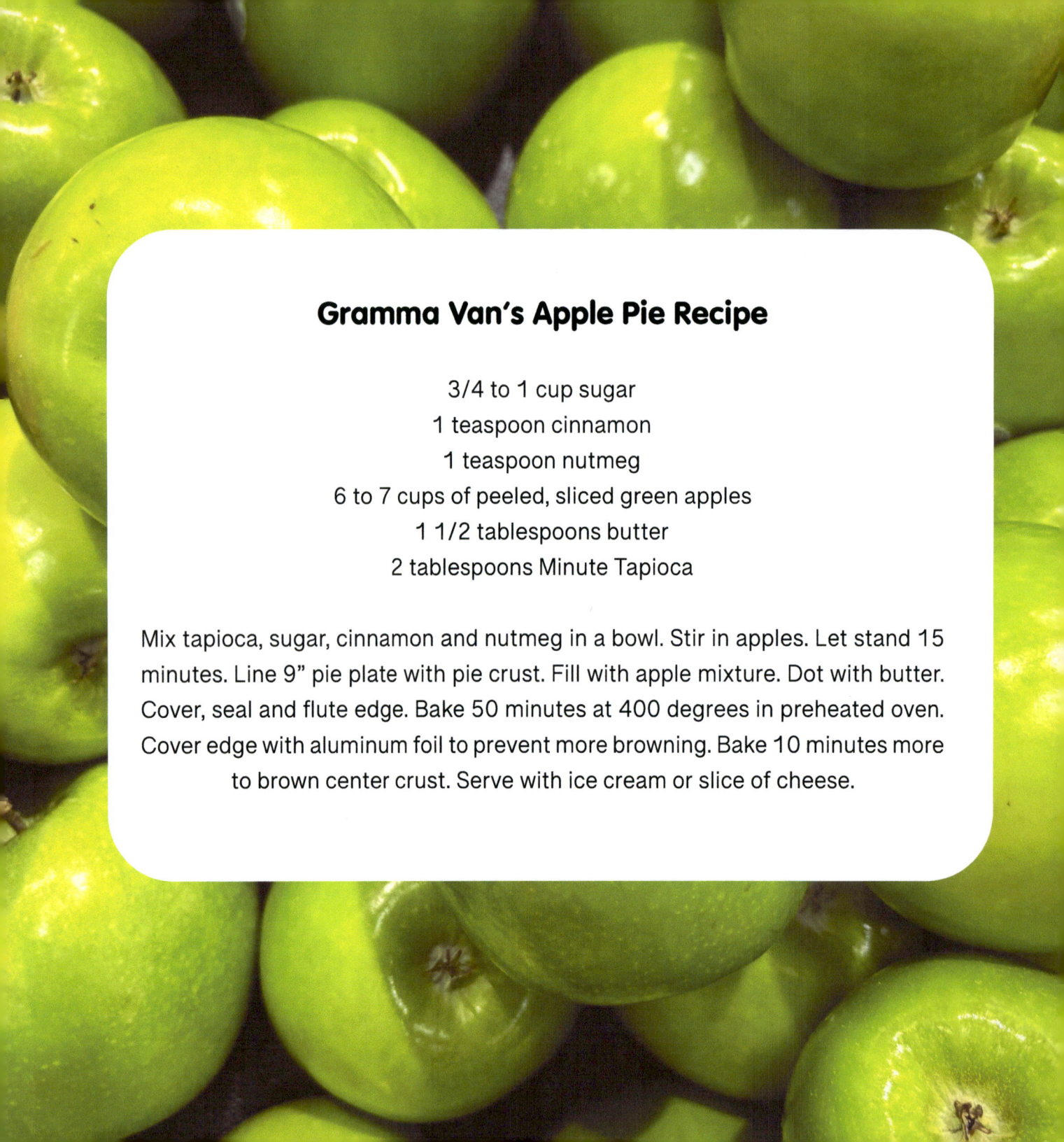

Gramma Van's Apple Pie Recipe

3/4 to 1 cup sugar
1 teaspoon cinnamon
1 teaspoon nutmeg
6 to 7 cups of peeled, sliced green apples
1 1/2 tablespoons butter
2 tablespoons Minute Tapioca

Mix tapioca, sugar, cinnamon and nutmeg in a bowl. Stir in apples. Let stand 15 minutes. Line 9" pie plate with pie crust. Fill with apple mixture. Dot with butter. Cover, seal and flute edge. Bake 50 minutes at 400 degrees in preheated oven. Cover edge with aluminum foil to prevent more browning. Bake 10 minutes more to brown center crust. Serve with ice cream or slice of cheese.

Special Thanks

Jadon, Grampa Ted, Larkin and Britt Sekulic

popstradamus.com

Paperback ISBN: 979-8-9916273-6-8

Book design by Britt Sekulic

© 2025 Copyright Carol Van Zanden. All rights reserved.

About the Author

Carol Van Zanden is a retired Home Economics and kindergarten teacher who has lived in the Pacific Northwest all her life. With her BA and MA in Education, her 38-year professional career encompassed teaching in early elementary, high school, college, and adult education. She and her husband, Ted, raised their children in Oak Harbor, Washington.

Through the years, Carol combined her love of family and photography, capturing memories of their grandchildren's visits with her camera. Years ago, she cut and pasted a series of books together using photos of her granddaughter and grandson to be read to by their parents and written in early childhood language for them to read by themselves. Not knowing that someday her prayers would be answered, with the help of self-publishing and collaborating with a local bookmaker and friend, the books would be brought to life professionally for other parents to read to their children and for them to read themselves.

Did you enjoy this book? Try another book in the series:

www.ingramcontent.com/pod-product-compliance
Lightning Source LLC
Chambersburg PA
CBHW061401090426
42743CB00002B/105